D0856058

WE CAN READ about NATURE!™

ANIMAL TALK

by Melissa McDaniel

***B*ENCHMARK *B*OOKS**

MARSHALL CAVENDISH
NEW YORK

With thanks to
Susan Jefferson, first grade teacher at Miamitown
Elementary, Ohio, for sharing her innovative teaching
techniques in the Fun with Phonics section.

Benchmark Books
Marshall Cavendish Corporation
99 White Plains Road
Tarrytown, New York 10591
Website: www.marshallcavendish.com

Text copyright © 2002 by Marshall Cavendish

Photo Research by Candlepants, Inc.

Cover Photo: *Photo Researchers, Inc.*, Renee Lynn

The photographs in this book are used by permission and through the courtesy of:
Corbis: Jeffrey L. Rotman, 4; Kevin Schafer, 5. *Photo Researchers:* Jerry L. Ferrara, 6;
Gregory G. Dimijian, 7 (top); S.R. Maglione, 7 (bottom); Dan Guravich, 10; Jeanne
White, 11; Carl Purcell, 12; Tim Davies, 13, 22 (bottom), 25 (top), 25 (bottom); Art
Wolf, 14; M.H. Sharp, 18; Harry Rogers, 19; F. Stuart Westmorland, 21; C.K. Lorenz, 22
(right); Renee Lynn, 24; M. Boet/A. Jacana, 27 (top); Jeff Lepore, 27 (bottom); Wolfgang
Kaehler, 29. *Animals Animals:* Gerard Lacz, 8, 20; George Bernard, 9; Gregory Brown,
15; Maresa Pryor, 16; Erwin & Peggy Bauer, 17; Joe McDonald, 23; Richard La Val, 26;
Johnny Johnson, 28.

Library of Congress Cataloging-in-Publication Data

McDaniel, Melissa.
Animal talk / by Melissa McDaniel.
p. cm. — (We can read about nature)
Includes index (p.32).
ISBN 0-7614-1253-0
1. Animal communication—Juvenile literature. [1. Animal communication. 2.
Animals—Habits and behavior.] I. Title. II. Series.

QL776.N53   2001   591.59—dc21   00-068011

Printed in Italy

1 3 5 6 4 2

Look for us inside this book.

ant

buck

chimpanzee

deer

dolphin

elephant

fox

giraffe

gorilla

lion

penguin

wolf

People say hello in many ways.

Bedouins greet each other in Egypt.

And so do animals.

Animals say hello with a kiss or a nip.

Elephants wrap trunks.

Lions rub
heads.

Tall or small, animals ask, "Who are you?"

Giraffes rub necks.

Ants touch feelers.

Animals use all their senses to talk. They look and listen.

Walruses

An Airedale dog

11

They touch and taste.

Zebras

Lions

They even use their sense of smell. "Who's been here?" asks the fox.

A red fox

This buck is rubbing a tree to leave his scent.

"This is my place," he says.

A white-tailed deer

What else do animals say? "Danger!" warns the deer with a flick of its tail.

A white-tailed deer

"Stay away from here," howls the wolf.

A gray wolf

"Look at me!" says the bird.

A great egret

"Back off," says the bug.

A praying mantis

Dolphins use sounds to talk. They whistle underwater.

One makes a sound.
Another answers.

Bottlenose dolphins

One leaps. Another follows.

Apes use their faces to show how they feel.

An orangutan

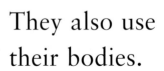

A chimpanzee

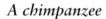

They also use their bodies.

22

"Don't mess with me," says
the gorilla.

A gorilla

Chimpanzees grunt or hoot to say, "Over here! There's food!"

They hug
to say,
"It's okay."

People have
taught some
apes to
make signs.

Other animals have their own ways of talking to people.

Different kinds of animals can talk to each other.

"Leave me alone," says the bird to the sheep.

"Who are you?" the dog asks the squirrel.

What do you think these penguins are saying?

29

fun with phonics

How do we become fluent readers? We interpret, or decode, the written word. Knowledge of phonics—the rules and patterns for pronouncing letters—is essential. When we come upon a word we cannot figure out by any other strategy, we need to sound out that word.

Here are some very effective tools to help early readers along their way. Use the "add-on" technique to sound out unknown words. Simply add one sound at a time, always pronouncing previous sounds. For instance, to sound out the word **cat**, first say **c**, then **c-a**, then **c-a-t**, and finally the entire word **cat**. Reading "chunks" of letters is another important skill. These are patterns of two or more letters that make one sound.

Words from this book appear below. The markings are clues to help children master phonics rules and patterns. All consonant sounds are circled. Single vowels are either long –, short ˘, or silent /. Have fun with phonics, and a fluent reader will emerge.

When the letter "a" is followed by the letter "y," the "y" is considered a vowel, and the "two-vowels-together" rule is used. When two vowels are together the first vowel is long and the second vowel is silent. Long vowels say the name of the vowel.

w ā y̸ s   ō k ā y̸   s t ā y̸   s ā y̸   ă w ā y̸

When two vowels are together, the first vowel is long and the second vowel is silent. Long vowels say the name of the vowel.

l ī ø n s   d ē e̸ r   t ā i̸ l   l ē a̸ p s   t r ē e̸

A short "u" word is the letter "u" sandwiched between two consonants.

g r ŭ n t   h ŭ g   b ŭ g   b ŭ c k

Words that end with an "e" can be called magic e words. The "e" is silent, and the vowel becomes a long vowel. Long vowels say the name of the vowel.

ū s e   t ā s t e   h ē r e   l ē a v e

ā p e s   t h ē s e

fun facts

- Each chimpanzee has its own hoot. Chimps can recognize each other just by hearing these hoots.
- Bees do a dance that tells other bees where to find food. Their movements show how far to fly and in what direction.
- Elephants make a deep rumbling sound that people cannot hear. But other elephants can hear this sound five miles away.
- A gorilla named Koko has been taught more than a thousand hand signs.

glossary/index

about the author

Melissa McDaniel is a writer and editor living in New York City. The author of more than a dozen books for young people, she has written on topics ranging from movies to ducks. She is an avid hiker and loves wandering through New York City parks, marveling at all the wild creatures that live there.